# The Sounds Around Us

by Kelly Gaffney

a Capstone company — publishers for children

*Engage Literacy* is published in the UK by Raintree.
Raintree is an imprint of Capstone Global Library Limited, a company incorporated in England and Wales
having its registered office at 264 Banbury Road, Oxford, OX2 7DY – Registered company number: 6695582

www.raintree.co.uk

Editorial credits
Gina Kammer, editor; Charmaine Whitman and Clare Webber, designers;
Pam Mitsakos, media researcher; Tori Abraham, production specialist

Image credits
Alamy: Werli Francois, 4; Capstone Press: Capstone Studio/Karon Dubke, 21; Getty Images: Cultura RM Exclusive/
Robyn Breen Shinn, 9; Newscom: Rauschenbach, F./picture alliance/Arco Images G, 19; Science Source: Sophie
Jacopin, 11; Shutterstock: adriaticfoto, 14, AJP, 1, 22 top right, ananaline, 23 bottom left, Andrea Izzotti, 18, anigoweb,
design element, Catalin Petolea, 16, cbpix, cover bottom right, Chromakey, cover middle left, Darren Baker, 8, Dmitry
Kalinovsky, cover top left, esfera, cover bottom left, Faraways, 5, Francesca McConnell, cover middle background,
hwongcc, 22 bottom left, khuruzero, cover top right, kryzho , 7, Mark Herreid, 12, Mark Skalny, 17, Patricia Marks
back cover, Photo Melon, 22 bottom right, Trofimov Denis, 10, Viorel Sima, 23 top, Yuganov Konstantin, 6; SuperStock:
LAURENT/PASCAL/BSIP, 15; Thinkstock: Jupiterimages, 13

10 9 8 7 6 5 4 3 2 1
Printed and bound in China.

The Sounds Around Us

ISBN: 978 1 4747 3165 2

# Contents

Sounds and the world..................................4

Sounds that help us ...............................6

Sound and feelings ................................8

How do we hear sounds? ........................10

Different kinds of sounds.........................12

People who can't hear.............................14

Interesting things that sound can do ........16

Sounds that only animals can hear ..........18

Making sounds ....................................20

Glossary...............................................24

Index ...................................................24

# Sounds and the world

Stop for a moment and listen. What can you hear?

Perhaps you can hear the wind blowing through the trees or a bird chirping outside. There are sounds all around us, but we don't always notice them.

Hearing is one of our *senses*. The sounds we hear help us to understand what is going on around us.

Sounds help us to share thoughts and ideas with other people. Have you ever thought about all the different sounds you can make?

When we talk sounds come out of our mouth. We also make sounds when we laugh, cry or scream. Sometimes people clap their hands to *cheer* on their favourite team. The sounds we make often let other people know how we are feeling.

# Sounds that help us

The sounds we hear let us know what is going on nearby. Even if we are busy, some sounds will make us look up to see what is happening. A school bell might tell you that it is time to go to your lesson. The alarm on a clock can tell you that it is time to get out of bed.

Sounds can also warn us of danger. A fire alarm can let us know that there is a fire. The alarm warns us, even if we can't see or smell the fire.

The sound of a car's horn can also warn us. It can let us know that a car is nearby, even if we can't see it.

Animals make sounds that warn us, too. If you hear a dog growl, you know that the dog isn't happy and might bite.

# Sound and feelings

Sounds can even change the way we feel. Listening to music can make you feel happy, sad or like you want to get up and dance! A very loud crash can make you feel frightened. Sounds that go on and on for a long time can bother you. And the pitter-patter of rain on the window can make you feel relaxed.

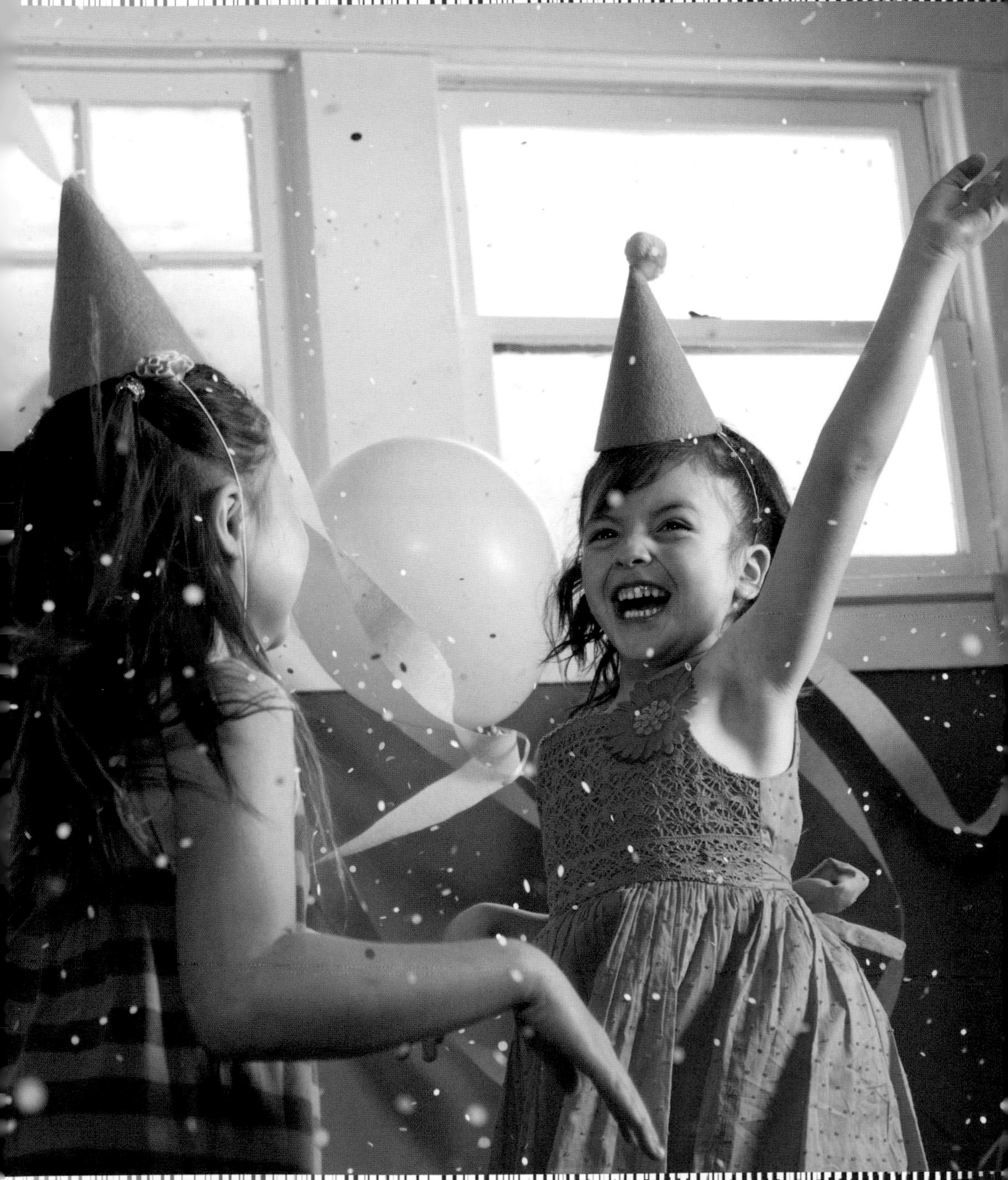

# How do we hear sounds?

All sound begins with a *vibration*. A vibration is when something moves back and forth very quickly. Sounds are made by these vibrations, which move in *waves*, up and down smoothly. We call these *sound waves*.

When you pluck a string on a guitar, you can see the string vibrate. It is the string vibrating that makes the sound. When the string stops moving, the sound stops.

## DID YOU KNOW?

You can feel electric speakers move when they make really loud sounds because they are vibrating.

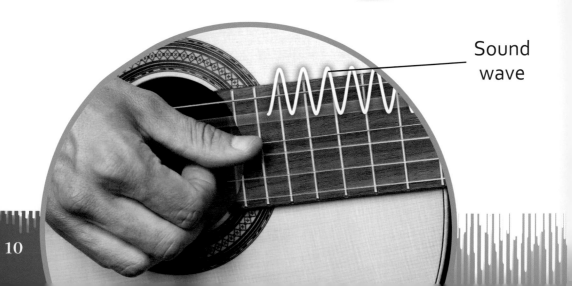

Sound wave

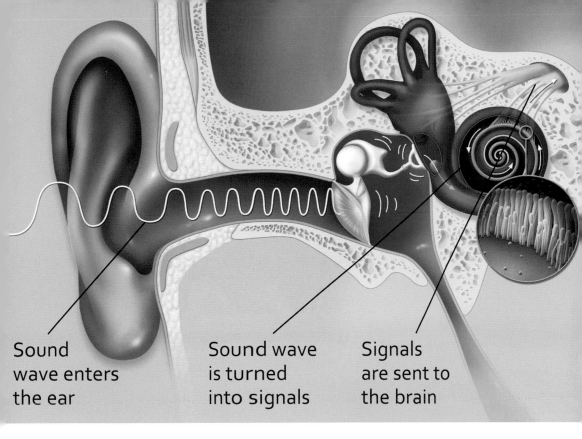

Sound wave enters the ear

Sound wave is turned into signals

Signals are sent to the brain

We hear sounds when sound waves go into our ears. Once sound waves are inside our ears, they are turned into messages, or signals. These messages are sent to the *brain*. When the message reaches our brain, we hear the sound. This all happens very quickly.

# Different kinds of sounds

The sounds we hear can be very loud or quiet. A loud sound is made by a big vibration. A quieter sound is made by a small vibration. If you hit a drum very hard, you make a big vibration, which makes a loud sound. If you tap the drum very gently, you make a very small vibration, which makes a very quiet sound.

Sounds can also be high or low. This is called *pitch*. An example of a high-pitched sound is a whistle. An example of a low-pitched sound is a rumble of thunder.

# People who can't hear

Some people can't hear sounds. Other people can hear some sounds but not very well. Many people who find it difficult to hear watch other people's mouths when they talk. This is called *lip-reading*. Some people use *sign language* to share ideas and thoughts with others.

**DID YOU KNOW?**

Everything that can be said with words can also be said with sign language.

A *hearing aid* is a special machine that can help some people hear. A hearing aid can make sounds louder. It can also make it easier to hear what someone is saying when there is a lot of other noise.

# Interesting things that sound can do

Sometimes sounds can do strange things. Have you ever called out and heard your own voice come back to you? This is called an *echo*. An echo happens when sounds hit something hard and bounce back.

You might hear an echo if you call out in a big empty building. You might hear an echo when you call out in a cave, too. This is because the sounds made by your voice are bouncing back to you.

→ man's voice
← echo

# Sounds that only animals can hear

Dolphins make some sounds that we can't hear. The sounds they make bounce off fish in the water around them. This tells the dolphin where the fish are.

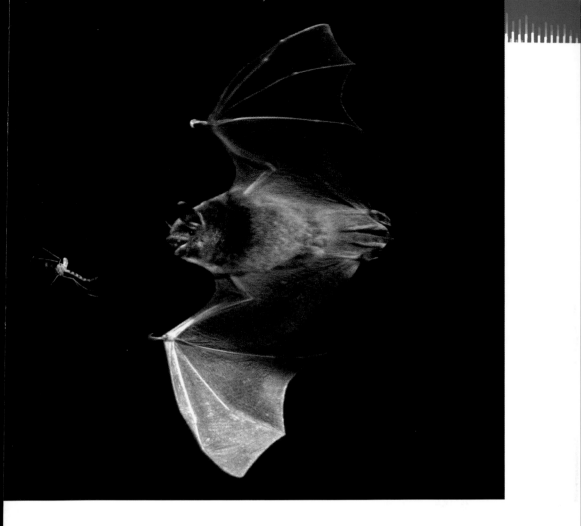

Bats also make some sounds we can't hear. They use these sounds to hunt for insects at night. The bat makes sounds that bounce off insects as they fly through the air. This helps the bat to find insects in the dark.

# Making sounds

You can learn about sound and even make your own music! You will need five glass bottles or drinking glasses that are the same size. You will also need a spoon or a pencil and a jug of water.

Put all the bottles next to each other on a table. Leave one bottle empty and fill the other four bottles with different amounts of water. Tap the spoon or pencil on the side of each bottle.

When you tap the bottle with the spoon or pencil, it makes the glass vibrate. This vibration makes a sound. The sound moves through the water in the bottle. Do all of your bottles make the same sound? Try to make some music by tapping the bottles.

Sounds are all around us. Stop and listen. What sounds do you hear? How do these sounds make you feel and what can they tell you about the world around you?

# Glossary

*brain* body part inside your head that controls your movements, thoughts and feelings

*cheer* shout to support someone

*echo* sound that returns after a travelling sound hits an object

*hearing aid* small electronic machine that people wear in or behind one or both ears; hearing aids make sounds louder

*lip-reading* watching someone's lips while the person is talking in order to understand what the person is saying

*pitch* how high or low a sound is

*sense* way of knowing about your surroundings; hearing, smell, touch, taste and sight are the five senses

*sign language* hand signs that stand for words, letters and numbers

*sound wave* wave or vibration that can be heard

*vibration* fast movement back and forth

*wave* something that follows a curving, up and down pattern

# Index

echoes   16, 17

feelings   5, 8, 10, 22

hearing   4, 6, 7, 11, 12, 14, 15, 17, 18, 19, 22

hearing aids   15

pitches   13

sign language   14

talking   5, 14

vibrations   10, 12, 20

voice   16–17

warnings   7